Mahmood Yekeh Yazdandoost

# Research for Innovation

Mahmood Yekeh Yazdandoost

# Research for Innovation

## Resilient Thinking

Noor Publishing

**Imprint**
Any brand names and product names mentioned in this book are subject to trademark, brand or patent protection and are trademarks or registered trademarks of their respective holders. The use of brand names, product names, common names, trade names, product descriptions etc. even without a particular marking in this work is in no way to be construed to mean that such names may be regarded as unrestricted in respect of trademark and brand protection legislation and could thus be used by anyone.

Cover image: www.ingimage.com

Publisher:
Noor Publishing
is a trademark of
Dodo Books Indian Ocean Ltd. and OmniScriptum S.R.L publishing group

120 High Road, East Finchley, London, N2 9ED, United Kingdom
Str. Armeneasca 28/1, office 1, Chisinau MD-2012, Republic of Moldova, Europe
Managing Directors: Ieva Konstantinova, Victoria Ursu
info@omniscriptum.com

Printed at: see last page
**ISBN: 978-620-9-13669-6**

# *Research for Innovation*

***By:***

***Mahmood Yekeh Yazdandoost***

***Environment is all about connection and all making connection.***

***Are we doing our best at what we do? Perhaps the difference between just doing and doing our best is the difference that solves most of the problems around us.***

***Sustainable development from a psychological and philosophical perspective is: self-knowledge, community knowledge, species adaptation, ecosystem connections, and resilience to exist and live, and to continue; this path will continue until there is no obstacle in the way.***

***Sustainable development, an unfulfilled dream! This unfulfilled dream will end sadly the day, when nothing more can be done.***

# ***Table of Content:***

# Nature-based Resilient Infrastructure for Sustainable Urban Development

**BY:** Mahmood Yekeh Yazdandoost (PhD), Advisor Expert, Department of Environment (DoE), Iran.
yazdandoost3000@gmail.com

## Abstract:

Nature can donate important contributions in fortifying resilience to environment. Nature-based Resilient Infrastructure (N-bri) can be applied as a replacement or supplement to grey infrastructure, and furnish a variety of benefits and strengths to society, including economic boost and environmental conservation. In this regard, infrastructure play a crucial role to human well-being and good societal operation with functional economy. Cities encounter multi environmental and socio-economic challenges that in conjunction pressurizes the resilience of urban areas and communities. Nature-based solutions directly convey ways to increment urban resilience. N-bri can present a fundamental function to assist the implementation of NbS. This research article guide a framework that evaluates the approach of N-bri to NbS implementation. The result indicate while N-bri contribute significantly to NbS outcome, still there are unresolved gaps in unexpected environmental changes that impact sustainable urban development.

## Keywords:

Nature-based Resilient Infrastructure, Nature-based Solutions, Ecosystems Services Delivery, Ecosystem-based Approaches, Sustainable Development Goals.

## Introduction:

Nature can donate important contributions in fortifying resilience to environment. Nature-based Resilient Infrastructure (N-bri) can be applied as a replacement or supplement to grey infrastructure, and furnish a variety of benefits and strengths to society, including economic boost and environmental conservation. In this regard, infrastructure play a crucial role to human well-being and good societal operation with functional economy. N-bri is a term that embraces a broad range of environmental friendly offers. Through its compatibility with nature, sustainably protect natural ecosystems as well as urban areas that benefit nature and people together.

Nature-based Resilient Infrastructure (N-bri), rely on Nature-based Solutions (NbS) to effectively contribute to Sustainable Development Goals (SDGs) and the transformation to better environment. NbS is an integrated approach to undertake activities for societal challenges, climate change, biodiversity constraint, ecological well-being and sustainable development, in order to conserve and amend natural ecosystems, and improve urban engineered areas for the benefits of human and other living creatures (CBD, 2004; IUCN, 2020).

NbS accomplish at the landscape scale and Nbri rely on multiple network arrangement, landscape dissimilarity and connectivity (Allen et al., 2016; Messier et al., 2019; Mina et al., 2021). In landscape scale, connectivity, modularity and status of patch size may amend society resilience to disorders and enhance energy flows in landscapes to conserve biodiversity (Kool et al., 2013; Wintle et al., 2019).

Human health and sustainability is interlinked to ecosystem health, so for optimization of societal well-being a consolidated approach towards human ecosystem is required. Ecosystems deliver various tasks that supports living creatures, including humans and their surrounding environment. Ecosystem-based

Approaches (EbA) came forth for practical solutions as NbS to address socio-environmental challenges like, global warming, climate change, natural catastrophe, water scarcity, health, safety and development. NbS through Ecosystems Services Delivery (ESD) buffer cities and increase their resilience to climate change and natural disasters, and being effectively resilient and efficient. In this regard, NbS tackle societal problems through ESD (Cohen-Shacham et al., 2016). Moreover, Gomez-Baggethun et al. (2013) used the term "ecosystem services" in urban areas for understanding the resilience capacity in cities. For empowering urban areas to withstand calamities and be able to adapt to unexpected changes, the concept of resilience and their path of action initially should be realized (Pickett et al., 2004; Bush and Doyon, 2019).

As it is clear, cities encounter multi environmental and socio-economic challenges that in conjunction pressurizes the resilience of urban areas and communities. Nature-based solutions directly convey ways to increment urban resilience. N-bri can present a fundamental function to assist the implementation of NbS. This research paper guide a framework that evaluates the approach of N-bri to NbS implementation and demonstrate how N-bri supports sustainable development programs. The result indicate while N-bri contribute significantly to NbS outcome, still there are unresolved gaps in unexpected environmental changes that impact sustainable urban development.

## Methodology:

In methodology, the indicators such as; carbon dioxide reduction, water resources handling, air pollution absorption and reduction, air quality condition, rain runoff constraint, urban heat reduction, energy improvement, biodiversity protection and health promotion have been taken into consideration.

## Geographical area:

Tehran is the major city and capital of Iran. The city is the ministerial and administrative center with a population of above 9.8 million in the city according to 2025 censes, and 16.8 million in the metropolitan premises. It is the most densely populated metropolitan in Iran and Western Asia, the second-largest metropolitan area in the Middle East after Cairo, and the 24th most heavily populated metropolitan area in the world.

## Results and Discussion:

The success of NbS will be characterized by its resilient capacity to withstand environmental disturbances, as it is evident in the adaptation process to global climate change (Key et al., 2022; Turner et al., 2022). N-bri can be recognized for its capacity to assist climate change adaptation and mitigation program. The concern is the ecological (natural and man-made) resilience of NbS with variety of protections and functions. Community engagement and their socio-ecological linkage should be driving force and an integral part of NbS (Calliari et al., 2022; Seddon et al., 2020; Woroniecki et al., 2023).

Resilience Infrastructure (RI) has notable consideration in urban policy development. Constructing Urban Resilience Infrastructure (URI) involve long-term Trajectory-based Approach (TbA) to urban arrangements and perspectives, for necessary transformational change (Elmqvist et al., 2014). In this regard, the connection between resilience and sustainability can assist useful trajectories for urban development (Elmqvist et al., 2019).

N-bri fulfill several functions to promote resilience capacity of metropolitan cities for present challenges and future uncertainties. Urban Resilient Infrastructure (URI) is enhanced via the incorporation of nature-based solutions in urban environment. NbS provide the basic condition for human interaction and solidarity, and contribute to physical, mental, social and spiritual health and prosperity (Jennings and Bamkole, 2019). NbS are decentralized and distributed systems of infrastructure which are more resilient than centralized conventional infrastructure in urban areas.

Response diversity is crucial for ecological system to stay resilient, because presents effective relief. Hence, resilience is sustained by responses of living organisms throughout landscapes (Mori et al., 2013). Urban forests is a fundamental fragment of resilient infrastructures that supply numerous ecosystem services to the social and economic welfares of societies. Besides providing ecological stability, rectify air quality and create good habitats for wild flora and fauna by securing food through pollination and maintaining biodiversity (Kong et al., 2021; Zhang et al., 2024).

There is a wide evidence that, the ecological connectivity in a landscape contribute to resilience capacity (Dudney et al., 2022; Shackelford et al., 2017). Also the modularity ability with firm connectivity at the local levels when combined with delicate connectivity at spacious scales determine the extent of resilience

(Daigle et al., 2020; Newman, 2003). The quantity of connectivity and modularity regulate ecosystem recuperation levels from disorders and perseverance in the landscape (Oliver et al., 2015; Walker, 2020).

Urban resilience consider as integrated approach in city planning to sustain cities from numerous shocks. It is by implementing proactive adaptation strategies to make sure that the health and well-being of people is at the center of consideration. The cities resilience capacity is based upon their capability to regulate conditions at the time of change (Alberti and Marzluff, 2004; Pickett et al., 2004). Resilience embrace reacting to the slow changes and persistent tensions of socioeconomic type, and the sudden shocks, associated with natural disasters (Resilient Melbourne, 2016).

NbS are significant instrument when administered to engineered environment to tackle multiple challenges of urban settings. The implementation of Nbri such as green technology provide energy efficiency and aesthetic characteristics to urban areas (Bayulken et al., 2020). The formation of green layout like urban forests and wetlands ameliorate urban terrain and reinforce urban resilience. In this way, nature can be resettled back in metropolitan cities through NbS delivery. Therefore, N-bri through NbS implementation would achieve regeneration to promote health and well-being in the society. Regrettably, the Tehran metropolitan with dense urban construction and high population density, do not have adequate green infrastructures. The conventional approach to urban expansion distinguished by large grey infrastructure and intense resource and communication practices, has elevated the vulnerability of urban cohesiveness to adverse environmental impact.

## Recommendations:

-Protect and restore natural and man-made ecosystems, through NbS development.

-Design, implement and monitor NbS in collaboration with indigenous people and local communities.

-Ecosystem health and connectivity should be the central core of NbS.

-Enhance urban health through greening policies and strategies.

-Increase the resilience capacity, in the built environment.

## Conclusion:

This research paper illustrated Nature-based resilient infrastructure for sustainable urban development via contribution of NbS in constructing urban resilience. Resilience provide a good opportunity for urban planners to develop cities green transformational change.

The significance of N-bri to the outcome of NbS in enhancing and protecting ecosystems connectivity and climate change adaptation to rehabilitating biodiversity, incrementing carbon sequestration and relieving drought strikes has become increasingly important for social benefits. N-bri may be defined as "the ecosystem (natural and man-made) capability to withstand perturbations without qualitatively shifting to a new condition, governed under new operation".

N-bri consider to be greater than retrieving, because it is the capacity of a system to adapt, unify and flourish during the time of change or disruption. N-bri accordingly is a trajectory-based approach towards desirable stability. Therefore through N-bri, we can maximize Urban Areas Resilient Capacity (UARC) and develop efficiency for present and future challenges.

Urban planning is principally anthropocentric in nature. Metropolis should be recognizable as a hospitable natural environment in urban planning. A novel multi-species attitude for the admission of nonhuman entity to our surrounding is required. To magnify the practicality of ecological systems in enhancing resilience, the ecosystem approach should be at the core of urban designing programs, which are adjusted by climate risk details, ecosystem appraisals and susceptibility surveys.

The demand for greening infrastructure planning has emerged as a prerequisite for mitigating climate change, promoting health and managing water resources. Therefore, NbS can bring into services the utilization of ecosystems and biodiversity for climate change mitigation and adaptation, increase the resilience capacity, and promote the population health and well-being in a metropolitan city like Tehran. The N-bri can stand as a functional instrument to merge greening strategies into urban amendment operations. However the utilization of NbS at present is confined.

NbS are an executable scheme to consolidate the resilience of societies by promoting the ecological systems well-functioning and furnishing socio-ecological as well as economic benefits to people. Urban ecosystems if planned properly can address substantial role in soil and water management and regulate effectively atmospheric temperature. Policy-makers should take into consideration the territorial characteristics, as solution that work well in one urban area may not work effectively in other urban areas. Moreover, urban development planning should considerably fortify the adaptive capacity of urban areas and transform cities to resilient standing.

As a result, Nature-based Resilient Infrastructure (N-bri) can constitute a noteworthy ready for use capacity to urban areas in response of vulnerability to climate change. Extensive acquisition of N-bri in urban structure needs better institutional communication and improved financial backstop to support suitable NbS intervention.

## Acknowledgement:

All the scientific literatures released by academia, UN and other agencies including Department of Environment that have been used in this research paper are highly appreciated.

## References:

-Alberti, M., and Marzluff, J. M. (2004). Ecological resilience in urban ecosystems: Linking urban patterns to human and ecological functions. Urban Ecosystems, 7(3), 241–265.

-Allen, C. R., et al. (2016). Quantifying spatial resilience. Journal of Applied Ecology, 53, 625–635.

-Bayulken, B., et al. (2020). How are nature based solutions helping in the greening of cities in the context of crises such as climate change and pandemics? A comprehensive review. J. Clean. Prod. 2020, 288, 125569.

-Bush, J. and Doyon, A. (2019). Building urban resilience with nature-based solutions: How can urban planning contribute? Cities 2019, 95, 102483.

-Calliari, E., et al. (2022). Building climate resilience through nature-based solutions in Europe: A review of enabling knowledge, finance and governance frameworks. Climate Risk Management, 37, 100450.

-Cohen-Shacham, E., et al. (2016). Nature-based solutions to address global societal challenges. Gland, Switzerland.

-Convention on Biological Diversity-CBD (2004). The ecosystem approach (CBD guidelines). Secretariat of the Convention on Biological Diversity.

-Daigle, R. M., et al. (2020). Operationalizing ecological connectivity in spatial conservation planning with Marxan Connect. Methods in Ecology and Evolution, 11, 570–579.

-Dudney, J., et al. (2022). Capacity for change: Three core attributes of adaptive capacity that bolster restoration efficacy. Restoration Ecology, e13647.

-Elmqvist, T., et al. (2014). Exploring urban sustainability and resilience. In L. Pearson, P. Newton, & P. Roberts (Eds.). Resilient sustainable cities: A future (pp. 19–29). Abingdon: Routledge.

-Elmqvist, T., et al. (2019). Sustainability and resilience for transformation in the urban century. Nature Sustainability, 2(4), 267–273.

-Gomez-Baggethun, E., et al. (2013). Urban ecosystem services. In T. Elmqvist, M. Fragkias, J. Goodness, B. Güneralp, P. J. Marcotullio, R. I. McDonald, & C. Wilkinson (Eds.). Urbanization, biodiversity and ecosystem services: Challenges and opportunities. A global assessment (pp. 175–251). Dordrecht: Springer.

-International Union for Conservation of Nature-IUCN (2020). Global standard for nature-based solutions. A user-friendly framework for the verification, design and scaling up of NbS. https:// www. IUCN.

-Jennings, V., and Bamkole, O. (2019). The relationship between social cohesion and urban green space: An avenue for health promotion. International Journal of Environmental Research and Public Health, 16(3).

-Key, I. B., et al. (2022). Biodiversity outcomes of nature-based solutions for climate change adaptation:

Characterizing the evidence base. Frontiers in Environmental Science, 10, 905767.

-Kong, X., et al. (2021). Review on Urban Forests and Trees as Nature-Based Solutions over 5 Years. Forests 2021, 12, 1453.

-Kool, J. T., et al. (2013). Population connectivity: Recent advances and new perspectives. Landscape Ecology, 28, 165–185.

-Messier, C., et al. (2019). The functional complex network approach to foster forest resilience to global changes. Forest Ecosystems, 6, 1–16.

-Mina, M., et al. (2021). Network analysis can guide resilience-based management in forest landscapes under global change. Ecological Applications, 31, e2221.

-Mori, A. S., et al. (2013). Response diversity determines the resilience of ecosystems to environmental change. Biological Reviews, 88, 349–364.

-Newman, M. E. (2003). The structure and function of complex networks. Society for Industrial and Applied Mathematics Review, 45, 167–256.

-Oliver, T. H., et al. (2015). Biodiversity and resilience of ecosystem functions. Trends in Ecology & Evolution, 30, 673–684.

-Pickett, S. T. A., et al. (2004). Resilient cities: Meaning, models, and metaphor for integrating the ecological, socio-economic, and planning realms. Landscape and Urban Planning, 69(4), 369–384.

-Resilient Melbourne (2016). Resilient Melbourne: Resilience strategy for greater Melbourne. Melbourne: City of Melbourne.

-Seddon, N., et al. (2020). Understanding the value and limits of nature-based solutions to climate change and other global challenges. Philosophical Transactions of the Royal Society B, 375, 20190120.

-Shackelford, N., et al. (2017). Isolation predicts compositional change after discrete disturbances in a global meta-study. Ecography, 40, 1256–1266.

-Turner, B., et al. (2022). The role of nature-based solutions in supporting social-ecological resilience for climate change adaptation. Annual Review of Environment and Resources, 47, 123–148.

-Walker, B. H. (2020). Resilience: What it is and is not. Ecology and Society, 25(2), 11.

-Wintle, B. A., et al. (2019). Global synthesis of conservation studies reveals the importance of small habitat patches for biodiversity. Proceedings of the National Academy of Sciences of the United States of America, 116, 909–914.

-Woroniecki, S., et al. (2023). Contributions of nature-based solutions to reducing people's vulnerabilities to climate change across the rural Global South. Climate and Development, 15, 590–607.

-Zhang, P., et al. (2024). The importance of current and potential tree canopy on urban vacant lots for landscape connectivity. Urban For. Urban Green. 2024, 94, 128235.

# The Impact of Good Governance in Iran's Landscape Restoration and rehabilitation program

**BY:** Mahmood Yekeh Yazdandoost (PhD), Advisor Expert, Department of Environment (DoE), Iran
yazdandoost3000@gmail.com

## Abstract:

Landscape restoration is the long-term activity of retrieving ecological functionality, and raising people's health across degraded land. It concerns to bring back the biological productivity of a region in order to attain well-being for human and safety for the planet with all its creatures. Vast areas of Iran facing ecosystem degradation, causing biodiversity loss and its impact on humanity. Good governance approach is a decisive factor in the efficacy, sustainability and productivity of resources as well as conservation and rehabilitation of habitats and protected areas in Iran. Landscape restoration increases ecosystem functions and services, significantly spread proper supply over customary place-based standpoints of terrain management. Nevertheless in landscape restoration there are multiple obstacles to prevail. Good policies needed in order to provide extensive uptake from landscapes efficiency. Restoration of forests by avoiding the logging practices legally or illegally, furnish advantages like, climate change mitigation, while endemic species for restoration processes can restore various forestry ecological services. This research paper considers the opportunities to upgrade restoration and rehabilitation based on Governance, Policy and Technical (GPT) issues.


## Keywords:

Landscape, Ecosystem, Restoration, Rehabilitation, Sustainability, Biodiversity, Connectivity.


## Introduction:

Degradation and deforestation are main environmental challenges these times. Iran because of climate change and intrusions through industrial activities and development consider to be very impacted. Reversing land degradation especially in protected areas is the national priority for biodiversity conservation and ecosystem services delivery in Iran. For achieving the

Sustainable Development Goals (SDGs), executable actions are needed to restore degraded land. Landscape restoration is the long-term activity of retrieving ecological functionality, and raising people's health across degraded land. It concerns to bring back the biological productivity of a region in order to attain well-being for human and safety for the planet with all its creatures. Vast areas of Iran facing ecosystem degradation, causing biodiversity loss and its impact on humanity. Good governance approach is a decisive factor in the efficacy, sustainability and productivity of resources as well as conservation and rehabilitation of habitats and protected areas in Iran. Restoration considered to be a key element in Iran for promoting conservation priorities and principles. The Restoration Opportunities Assessment Methodology (ROAM) expected to increase connectivity between landscapes, ecosystems and habitats.

The reason for the extensive ecosystems degradation in Iran include: expansion of agricultural fields for more crop production, pasturing of the widespread livestock, conversion of forested land for road construction, uncontrolled mining and industrial development on virgin habitat. While Extensive Ecological Restoration (EER) has the capacity to be a fundamental constituent in integrative management of nature.

There is a growing confirmation that good governance is a basic key to a fair and sustainable protected areas conservation as well as habitat restoration (Bennett et al., 2019; Mansourian, 2016; Springer et al., 2021). This comprises conservation achievement that broadly contributed by indigenous and local communities (Dawson et al., 2021; IPBES, 2019). Restoration ameliorates soil stability and consistency, water quality, habitat regeneration and biodiversity growth, healthy ecosystem and sustainable development, climate secureness and stability, societal solidarity and recreational advantages and collaboration among stakeholders (Barral et al., 2015; Alexander et al., 2016). Ecological restoration may also furnish economic interests (Blignaut et al., 2014; ELD Initiative, 2015).

Landscape restoration increases ecosystem functions and services, significantly spread proper supply over customary place-based standpoints of terrain management. Nevertheless in landscape restoration there are multiple obstacles to prevail (Menz et al., 2013). Good policies needed in order to provide extensive uptake from landscapes efficiency.

Landscape restoration is a prominent and developing exercise throughout the globe. As the world population increases in number and the request for clean environment and ecosystem services rises up, the restoration program become inevitable. Layout of landscape restoration via integrated perspective consider a few important phases, including; applying grasses and tree mixtures to imitate ecological fabric and functions of natural environment, reversing the habitat's alterations, developing and connecting patches of ecosystems and expanding ecological network systems. Exerting these phases furnish substantial situation for restoration program in areas with high degree of degradation (Scherr and McNeely, 2008). Water yields get affected by land cover, and with an excessive forest coverage has a high capacity to reduce water runoff and increase infiltration in the soil refurbishing underground aquifers (Zhang et al., 2001). Planting perennials endemic species reduce soil erosion and enhance biodiversity in a given area (Haregeweyn et al., 2012; Gibson-Roy et al., 2010).

Restoration is a forward-looking action, concentrating on future direction and prediction of climate and climate change, land use, population structure, socio-economic alteration and species sphere move. As a result restoration should be instructed by future ecosystem pathways in fusion with present and past resemblance (Hobbs et al., 2009). Also, restoration should be adjusted to various concerns of landscapes (Shackelford et al., 2013), so it will satisfy the needs landscapes to deliver various ecosystem services (Seabrook et al., 2011; Sayer et al., 2013; Hobbs et al., 2014); including cultural and social assets (Petursdottir et al., 2013; Shackelford et al., 2013).

Restoration is an instrument to tackle ecosystem degradation. Restoration of forests by avoiding the logging practices legally or illegally, furnish advantages like, climate change mitigation, while endemic species for restoration processes can restore various forestry ecological services (West et al., 2014; Tesfaye et al., 2015 ). This research paper considers the opportunities to upgrade restoration and rehabilitation based on Governance, Policy and Technical (GPT) issues.

Medium and long term objectives to alleviate problems included:

-Increase productivity of degraded land by reversing degradation processes.

-Reduce water runoff by improving watershed management.

-Conserve biodiversity through ecological connectivity processes.

-Engage local communities in restoration processes.

## Methodology:

The methodology considers the opportunities to upgrade restoration and rehabilitation based on Governance, Policy and Technical (GPT) issues, and designate priority actions for habitats and protected areas.

## Geographical area:

Iran's protected areas.

## Results and Discussion:

Considerable ecosystem changes, in association with swift global climate change and rising human population with socioeconomic change, create ecosystems shift (Hobbs et al., 2014). Novel ecosystems that arise through biotic and abiotic alterations prevent species from adapting to ecosystems modifications (Ito et al., 2013). This is especially in ecosystems with high degrees of degradations are worse, because provides lack of resiliency, that further create ecosystems collapse to rapid changes, like climate change (Hobbs et al., 2009).

Restoration attempts, design in the level of landscape with intention of re-organizing ecological unification and biological well-being that could contribute to human health (Maginnis & Jackson, 2003; Sabogal et al., 2015).The main goal is to upgrade the ecological resilience capacity through connectivity program, hence species, while the region faces rapid climate change and its consequences, can freely move into more hospitable habitat. The second goal is to merge effective redundancy into landscapes for supporting numerous ecosystem services, in spite of the decline of a few, due to considerable disturbances (Elmqvist et al., 2003).

Restoration does not always consider tree planting, but also consist of other administered natural regeneracy, agroforestry, and modified land operation to incorporate utilization of the mosaics of lands. Prioritizing regions that merge high level of possibilities for social and environmental gains with high level of practicability for considerable size of restoration is a necessary tool to attain goals. Attempt to face climate change, improve biodiversity and livelihoods via landscape restoration is a segment of recuperation.

Hence, for planning favorable restoration scheme, it is crucial to be informed what sort and mixture of reactions can successfully address degradation processes, permitting socioeconomic and environmental variation. For commencing, a survey and evaluation recognized the problems, and set objectives to tackle it. However, the main challenge in the study area was how to increment restoration capabilities that to be cost-effective and give rise to considerable outcome for Iran's socioeconomic and environmental objectives.

Two major problems we recognized included:

-National parks and protected areas have been fragmented by considerable alteration.

-Water quantity and quality have declined by excessive run-off due to deforestation.

## Recommendations:

-Integrate good governance programs into the institutional and policy settings, considering knowledge-based approach and accountability regulations.

-Secure comprehensive procedures with constructive contribution from proper institutions and authorize entry of Indigenous and local communities.

-Assist various stakeholders' participation that positively impact dynamic activities.

-Respect various knowledge via multi-directional instruction and collaboration.

-Integrate fundamental governance components in cost-benefit examination and consider their multiple propagation.

-Adapt restoration to various landscape functions for gaining multiple ecosystem services.

-Direct management strategies toward natural environmental conservation mechanisms and funding partnerships.

## Conclusion:

Ecological systems are dynamic and go through frequent shift in structure, composition and function because of complicated actions and reactions among species competition, climate change and variability, and moreover ecological niches, where they reside. This creates species dispersion, reduction, extinction or otherwise adaption and continued survival to new situation.

Ecological restoration is the main focal action in enhancing the various functionality and services of landscapes. It is the process of helping ecosystems to regain from degradation and impairment. Besides increasing resilience to the upcoming jerks and shifts, in swift global shift, ecological restoration can develop various benefits, especially when embraced as sustainable land handling creativity in landscapes efficiency.

Prosperous landscape restoration is a vital front looking perspective, which amplify landscapes resiliency and optimize ecosystems functionality in order any future challenges occur. Across various interests achieve by restoration, carbon sequestration seems progressively more advantageous. Restoring degraded land can remarkably enhance level of carbon in the soil and vegetation. These carbon gains benefit economy and livelihoods of local and indigenous communities and consider to be an important solution to cool down climate change.

In examining the practicable zones for restoration, high considerations accepted for high level biodiversity areas. These areas are biodiversity rich areas that deliver high ecosystems services, and another consideration directed towards areas that are hospitable to endangered species. In ecosystems where habitats through degradation become fragmented, restoration generate connections to make possible species movement. In Iran, the protected areas that incorporate excellent prospects for socio-environmental interests with great restoration practicability is a crucial tool to attain the aspiring restoration attentiveness.


## Acknowledgement:

All the scientific literatures released by academia, UN and other agencies including Department of Environment that have been used in this research paper are highly appreciated.

## References:

-Alexander, S., et al. (2016). The relationship between ecological restoration and the ecosystem services concept. Ecology and Society 21.

-Barral, M. P., et al. (2015). Quantifying the impacts of ecological restoration on biodiversity and ecosystem services in agroecosystems: A global meta-analysis. Agriculture, Ecosystems & Environment 202:223-231.

-Bennet, N., et al. (2019). 'Local support for conservation is associated with perceptions of good governance, social impacts, and ecological effectiveness'.

-Blignaut, J., et al. (2014). The economics of restoration: looking back and leaping forward. Annals of the New York Academy of Sciences 1322:35-47.

-Dawson, N., et al. (2021). 'The role of Indigenous peoples and local communities in effective and equitable conservation'. Ecology and Society 26(3): 19.

-ELD Initiative (2015). The value of land: Prosperous lands and positive rewards through sustainable land management. GIZ, Bonn, Germany.

-Elmqvist, T., et al. (2003). Response diversity, ecosystem change, and resilience. Frontiers in Ecology and the Environment 1:488-494.

-Gibson-Roy, et al. (2010). Expanding horizons for herbaceous ecosystem restoration: the Grassy Groundcover Restoration Project. Ecological Management & Restoration 11:176-186.

-Haregeweyn, N., et al. (2012). Integrated Watershed Management as an Effective Approach to Curb Land Degradation: A Case Study of the Enabered Watershed in Northern Ethiopia. Environmental Management 50:1219-1233.

-Hobbs, R. J., et al. (2009). Novel ecosystems: implications for conservation and restoration. Trends in Ecology & Evolution 24:599-605.

-Hobbs, R. J., et al. (2014). Managing the whole landscape: historical, hybrid, and novel ecosystems. Frontiers in Ecology and the Environment 12:557-564.

-Intergovernmental Science-Policy Platform on Biodiversity and Ecosystem Services (IPBES) (2019). Global assessment report on biodiversity and ecosystem services. E.S. Brondizio, J. Settele, S. Díaz and H.T. Ngo (Eds.). Bonn, Germany: IPBES Secretariat.

-Ito, T. Y., et al. (2013). Fragmentation of the Habitat of Wild Ungulates by Anthropogenic Barriers in Mongolia. PLoS ONE 8:e56995.

-Maginnis, S., & Jackson, W. (2003). The Role of Planted Forests in Forest Landscape Restoration. World, March, 25–27.

-Mansourian, S. (2016). 'Understanding the Relationship between Governance and Forest Landscape Restoration'. Conservation and Society 14(3): 267–278.

-Menz, M. H. M., et al. (2013). Hurdles and Opportunities for Landscape-Scale Restoration. Science 339:526-527.

-Petursdottir, T., et al. (2013). An Evaluation of the Short-Term Progress of Restoration Combining Ecological Assessment and Public Perception. Restoration Ecology 21:75-85.

-Sabogal, C., et al. (2015). Forest and landscape restoration: concepts, approaches and challenges for implementation. Unasylva, 66(245), 03–10.

-Sayer, J., et al. (2013). Ten principles for a landscape approach to reconciling agriculture, conservation, and other competing land uses. Proceedings of the National Academy of Sciences 110:8349-8356.

-Scherr, S. J., and J. A. McNeely. 2008. Biodiversity conservation and agricultural sustainability: towards a new paradigm of eco-agriculture landscapes. Philosophical Transactions of the Royal Society B: Biological Sciences 363:477-494.

-Seabrook, L., et al. (2011). Restore, repair or reinvent: Options for sustainable landscapes in a changing climate. Landscape and Urban Planning 100:407-410.

-Shackelford, N., et al. (2013). Primed for Change: Developing Ecological Restoration for the 21st Century. Restoration Ecology 21:297-304.

-Springer, J., et al. (2021). The Natural Resource Governance Framework: Improving governance for equitable and effective conservation. Gland, Switzerland: IUCN.

-Tesfaye, M. A., et al. (2015). Selection of Tree Species and Soil Management for Simultaneous Fuelwood Production and Soil Rehabilitation in the Ethiopian Central Highlands. Land Degradation & Development 26:665-679.

-West, T. A. P., et al. (2014). Forest biomass recovery after conventional and reduced-impact logging in Amazonian Brazil. Forest Ecology and Management 314:59-63.

-Zhang, L., et al. (2001). Response of Mean Annual Evapotranspiration to Vegetation changes at Catchment Scale. Water Resources Research 37:701-708.

# EIAs and SEAs are essential instruments in the planning and development process for large-scale projects

## Introduction:

Environmental Impact Assessments (EIA) and Strategic Environmental Assessments (SEA) are crucial components in the planning and development process, particularly for large-scale projects. These evaluations ensure that the environmental significance of developments are fully recognized and considered before any planning authorization is permitted. This guide provides a comprehensive outline of how EIAs and SEAs are required in the development process, concentrating on their importance, methodology, and the function they play in enhancing sustainable development.

## Realizing EIA and SEA:

**Environmental Impact Assessment (EIA):** An EIA is a process that recognizes and evaluates the environmental effects of suggested developments before commitments are made. It encloses a broad scope of environmental factors, like biodiversity, water, air and soil quality, and cultural inheritance. The EIA process is project-specific and is normally needed for large-scale developments that have the capability to crucially impact the environment.

**Strategic Environmental Assessment (SEA):** Unlike the EIA, which is project-specific, the SEA is an extensive assessment that evaluates the environmental impacts of suggested plans, programs, and policies. SEAs are applied to fulfill environmental considerations into the preparation and acquisition of strategic resolutions that guide future development projects.

## The Role of EIA and SEA in Large-Scale Developments:

For large-scale developments, EIAs and SEAs play a decisive role in securing environmental sustainability and adoption with regulatory requirements. These assessments:

**Inform Decision-Making:** EIAs and SEAs furnish noteworthy information about the environmental outcomes of suggested developments, assisting decision-makers to consider different choices and alleviation measures.

**Hinder Environmental Detriment:** By recognizing possible environmental impacts quickly in the planning process, these assessments help to avoid or diminish negative consequences on the environment.

**Enhance Public Involvement:** Both EIAs and SEAs initiate public advice, permitting stakeholders and the mainstream public to communicate their opinions and interests about the environmental impacts of developments.

**Contribute to Sustainable Development:** By incorporating environmental considerations into the planning and decision-making process, EIAs and SEAs contribute to the attainment of sustainable development goals.

## The EIA Process:

**The EIA process in most cases require several key stages:**

**Screening:** Specifies whether a suggested project needs an EIA based on its size, nature, and possible environmental impacts.

**Scoping:** Recognizes the key environmental issues and the extent of the assessment needed.

**Impact Assessment:** Evaluates the possible environmental impacts of the suggested project, examining both negative and positive effects.

**Mitigation:** Suggests measures to avoid, diminish, or remunerate for detrimental environmental impacts.

**Reporting:** Generates an Environmental Impact Statement (EIS) that sums up the findings of the EIA and dispense them to the decision-making body.

**Review:** The EIS is reconsidered by the appropriate authority to convince it meets the needed standards and satisfactorily addresses the possible environmental impacts.

**Decision-Making:** The decision-making body reviews the EIS, along with other elements, to make an informed decision about the suggested development.

**Monitoring and Compliance:** Guarantees that mitigation measures are applied and the project abides with environmental regulations.

## The SEA Process:

**The SEA process is almost identical to the EIA but acts at a strategic level. It involves:**

**Setting the Condition and Objectives:** Sets up the environmental, social, and economic objectives that the plan, programme, or policy should reach.

**Developing Substitutes:** Inspects different directions to attain the objectives while minimizing environmental impacts.

**Evaluating the Effects:** Evaluates the environmental outcomes of each substitute.

**Consultation:** Engages with the public and stakeholders to assemble feedback on the environmental report and the suggested plan or programme.

**Monitoring:** Applies a monitoring programme to trace the environmental effects of the plan or programme and guarantee that mitigation measures are effective.

## Key Environmental Factors Observed in an EIA:

**Biodiversity:** This comprised the effects on local flora and fauna, habitats, and conserved species. The judgement looks at possible habitat loss, fragmentation, degradation, and the impact on biodiversity conservation.

**Water Resources:** The survey evaluates the impact on water quality and quantity, including surface as well as groundwater. It reflects on pollution risks, variation in hydrology, and the potential for flooding or water shortages.

**Air Quality:** This requires determining the emission of pollutants into the air, including dust, gases, and particulate matter. The impact on local air quality and contribution to extensive air pollution issues are assessed.

**Soil and Land Use:** The EIA inspects the impact on soil quality and land use, including soil pollution, erosion, and the loss of productive land or agricultural fertility.

**Noise and Vibration:** The estimation considers the potential for increased noise and vibration levels, which can exert influence on human populations as well as wildlife.

**Landscape and Visible Impact:** This includes variations to the landscape and visual environment, recognizing how the development might change the nature and features of an area.

**Cultural Legacy:** The EIA measures the possible impacts on archaeological locations, historic structures, and cultural landscapes, considering direct as well as indirect effects.

**Climate Change:** Evaluations progressively reflect on a project's allotment to climate change via greenhouse gas emissions, and its vulnerability to climate change impacts, such as sea-level rise or enhanced weather variability.

**Socio-economic Impacts:** However not firmly environmental, socio-economic impacts, including effects on local communities, employment, and citizens health, are frequently considered due to their interrelation with environmental factors.

## The Role of Public Involvement in the EIA and SEA Processes:

Public involvement is a foundation of the EIA and SEA processes, ensuring that the perspectives and wisdom of the public and stakeholders are considered. This engagement contributes in a number of key directions:

**Enhancing Transparency and Trust:** Boosting Transparency and Confidence: Public participation assists to make the evaluation process more transparent and strengthens confidence between developers, officials, and the community.

**Upgrading Quality of Assessments:** Input from the public can deliver additional data and local knowledge that might not be in other respects obtainable, leading to more comprehensive and precise assessments.

**Recognizing Concerns and Other Possibilities:** Engagement with the community and stakeholders can focus attention on concerns and possible impacts that may not have been at first considered. It can also help in recognizing different perspectives or mitigation measures.

**Assisting Social Acceptance:** Through involving the public in the decision-making process, there is a considerable probability of social acceptance of the project, reducing the potential for disputes and postpones.

**Encouraging Environmental Supervision:** Public involvement motivates a sense of acquisition and accountability towards the local environment, enhancing sustainable development exercises.

## Conclusion:

EIAs and SEAs are essential instruments in the planning and development process for large-scale projects. They secure that environmental attentions are central to decision-making, enhancing sustainable development and inhibiting environmental depreciation. By following the structured processes of EIAs and SEAs, developers can alleviate environmental impacts, promote project sustainability, and contribute to the comprehensive goal of environmental safety.

## References:

-Abaza H., et al. (2004). Environmental Impact Assessment and Strategic Environmental Assessment: Towards an Integrated Approach. Economics and Trade Branch, UNEP, Geneva.

-Fischer, T. B. and Gonzalez, A. (2021). Handbook on Strategic Environmental Assessment. Cheltenham: Edward Elgar.

-Jha-Thakur, U. et al. (2009). Effectiveness of Strategic Environmental Assessment – The Significance of Learning. Impact Assessment and Project Appraisal 27(2): 133-144.

-Sadler, B. et al. (2011). Handbook of Strategic Environmental Assessment. London: Earthscan.

-Tetlow, M.F. and Hanusch M. (2012). Strategic Environmental Assessment: The State of the Art. Impact Assessment and Project Appraisal 30(1):15–24.

## Innovative thoughts for environmental footprint reduction

In recent years, the environmental practitioners become more focused on societal cohesiveness and solidarity to protect the environment. In this regard technology can play an important role in generating green energy, communication, transportation and waste reduction. However, we have to look back for grasping perspective on environment. Technology proceed real assurance in conserving human environment. “Green hydrogen” is becoming a crucial component of clean energy blend. The challenges are the high cost and important safety issues connected with generating hydrogen energy. Progresses in electrolyzer and fuel cell technology are confining the expenditure gaps. Also, advances in sensor technology is making safer to produce, transport and utilize hydrogen energy in combustion as well as electrochemical processes.

The idea of carbon footprint, refers to the total amount of greenhouse gases produced by human activity. These gases hold captive heat in the atmosphere, creating problem of global warming and climate change. Innovation in carbon footprint reduction is all about developing and executing novel vision, technologies, and procedures that limit greenhouse gas emissions. The straightforward direction for reducing carbon footprint is via technologies and procedures that limit energy consumption.  Improved energy generation directly defines limited fossil fuels use and reduction in greenhouse gas emissions. At the basic level, carbon footprint reduction initiates our perception about where emissions come from and how to execute strategies to reduce them. This comprises utilizing renewable energy sources, and building more sustainable options in our routine activities. Innovation act as a core zone in producing strategies to be workable and scalable across a range of systems.

As a result, innovation is necessary for carbon footprint reduction because would make provisions for implementing mitigation on greenhouse gas emissions across various sectors. From energy efficiency to renewable energy and sustainable exercises, innovation empowers us to generate a more environmentally sustainable and accountable future.

Energy efficiency is often considered as the initial support because it illustrates the most cost-productive process to carbon emissions reduction. Innovations in this domain comprises smart building technologies.

Industrial operations are also more effective by the use of improved sensors and data analytics that optimizes energy usage. Furthermore, innovations in material science proceeds to the evolution of lighter and tougher materials. Carbon capture and storage (CCS) technologies prevent carbon dioxide emissions from entering the atmosphere. It has the potential to remarkably reduce emissions from industrial provisions. The transportation sector is also an important contributor to greenhouse gas emissions. Innovations in this domain by developing cleaner and efficient transportation create options, like development of electric vehicles and hybrid vehicles.

Beyond technological progressions, innovations in policy and economic act as an essential part in carbon footprints reduction. Carbon pricing mechanisms, incentivizes trade to decrease their emissions by placing price on carbon. Government tax credits can also assist the acquisition of energy-efficient technologies. Evaluating the impact of innovation on carbon footprints needs an extensive perspective that investigates both direct as well as indirect effects. Life cycle assessments can measure the whole carbon footprint of a product. Econometric models can estimate the impact of policy interventions throughout emissions levels.

In spite of the noteworthy potential of innovation to carbon footprints reduction, a number of challenges and obstacles can impede its improvement. These contain high beforehand costs, lack of public perception, regulatory barriers, and disagreement from violated industries. Succeeding these obstacles needs a collaborative attempt from governments, trades and individuals. Public-private participations can assist to fund and catalyze the development and spread of innovative technologies. Instruction and expansion schemes can elevate consciousness of the interests of carbon footprint reductions and inspire the acquiring of sustainable exercises.

From an academic viewpoint, attending the carbon footprint reduction needs a systems-thinking attitude, admitting the interconnectedness of numerous sectors and operations that present to greenhouse gas emissions. Systems approach identifies that interventions in one sector can have cascading effects in other sectors. The interdisciplinary nature of carbon footprint reduction involves scrutinizing cross-sectoral impacts and recognizing solidarities. For example, sustainable land management strategies can increase carbon sequestration in soils while at the same time ameliorating agricultural fertility. Technological innovation presents a critical role in the move to a low-carbon economy. Advance technologies can considerably upgrade the performance and cost-benefit of renewable energy systems. These progresses need significant investment in research and development, and encouraging policy structure to motivate acquisition and scaling.

From an economic point of view, carbon footprints reduction implies producing market motivations that put in order economic pursuits with environmental desires. Carbon pricing mechanisms are planned to internalize the external costs of carbon emissions, constructing polluters pay for the environmental harm they bring about. These policies can push innovation by fabricating markets for low-carbon technologies and executions.

Carbon footprints reduction also necessitates directing social and behavioral features. Replacing consumer behavior and enhancing sustainable way of life are necessary for accomplishing remarkable emissions reductions. Also, social standards and ethical merits play important figures in framing behavior.

Moral principles are central to the argument over carbon footprint reduction. The impacts of climate change are inordinately felt by susceptible populations and developing countries, bringing up questions of environmental justice. Policies to reduce carbon emissions must be planned to elude deteriorating available inconsistencies and to secure that the benefits of a low-carbon economy are distributed in a fair and impartial manner.

Evaluating the successfulness of various carbon reduction strategies needs advanced modeling and estimation methodologies. Integrated assessment models can imitate the complex interconnections between the economy, energy structure and climate, permitting policymakers to assess the impacts of different policy frameworks. The succeeding course of innovation in carbon footprint reduction is probably to be shaped by different elements. These incorporate the consecutive development of renewable energy technologies, progresses in energy storage, the electrification of transportation, and the acquisition of circular economy principles. Moreover, artificial intelligence and big data analytics have the potential to accelerate innovation and bring sustainability solutions.

## References:

-Kim, Y.J. and C. Wilson (2019). Analyzing energy innovation portfolios from a systemic perspective. Energy Policy, Vol. 134, https://doi.org/10.1016/j.enpol.2019.110942.

-Kittner, N. et al. (2017). Energy storage deployment and innovation for the clean energy transition. Nature Energy, Vol. 2, https://doi.org/10.1038/nenergy.2017.125.

-Mazzucato, M. (2018). Mission-oriented innovation policies: Challenges and opportunities. Industrial and Corporate Change, Vol. 27/5, pp. 803-815, https://doi.org/10.1093/icc/dty034.

-Nemet, G.F. (2019). How Solar Energy Became Cheap: A Model for Low-Carbon Innovation. Routledge, London, https://doi.org/10.4324/9780367136604.

-OECD (Organization for Economic Co-operation and Development) (2015). Emerging technologies & firm dynamics: The implications of green growth. OECD, Paris, https://issue.com/oecd.publishing/docs/ggsd_2015_issuepaper2_finalpdf.

-Roberts, C. and F.W. Geels (2018). Political acceleration of socio-technical transitions. In: Jenkins, K.E.H. and D. Hopkins (eds.), Transitions in Energy Efficiency and Demand: The Emergence, Diffusion and Impact of Low-Carbon Innovation. Routledge, pp. 177-194, https://doi.org/10.4324/9781351127264.

-Wilcox, J. (2020). An electro-swing approach. Nature Energy, Vol. 5, pp. 121-122, https://doi.org/10.1038/s41560-020-0554-4.

-Wilson, C. and Y.J. Kim (2019). Analyzing energy innovation portfolios from a systemic perspective. Energy Policy, Vol. 134, https://doi.org/10.1016/j.enpol.2019.110942.

-Zemships (2008). One hundred passengers and zero emissions: The first ever passenger vessel to sail propelled by fuel cells. Brochure, Zemships, Hamburg, Germany, https://ec.europa.eu/environment/life/project/Projects/index.cfm?fuseaction=home.showFile&rep=file&fil=Zemships_Brochure_EN.pdf.

FSC
www.fsc.org
MIX
Papier | Fördert gute Waldnutzung
FSC® C083411

Kolibri 360 GmbH
Ferdinand-Jühlke-Straße 7
99095 Erfurt, Deutschland
produktsicherheit@kolibri360.de

Druck:
CPI Druckdienstleistungen GmbH
im Auftrag der
Kolibri 360 GmbH
Ferdinand-Jühlke-Str. 7
99095 Erfurt